I0468844

The Energy Secret

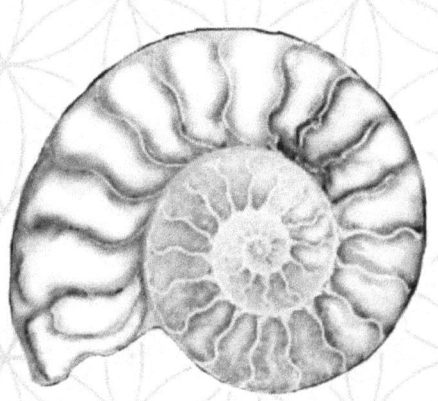

Karen Karasz

C. Copyright 2016 Karen Karasz. All rights
Reserved.
ISBN-13:
978-1530941469

Hello
there

I would like to share with you a secret...

A secret about the Energy of the Universe!

This is the flower of life.

It represents the Energy of the entire Universe!

In the beginning, there was sound.

Ohmmm mmmm...

............••

RESONANCE

Different sounds create different vibrational patterns.

Here are a few of my favorite patterns:

These patterns create a framework for the Energy to follow.

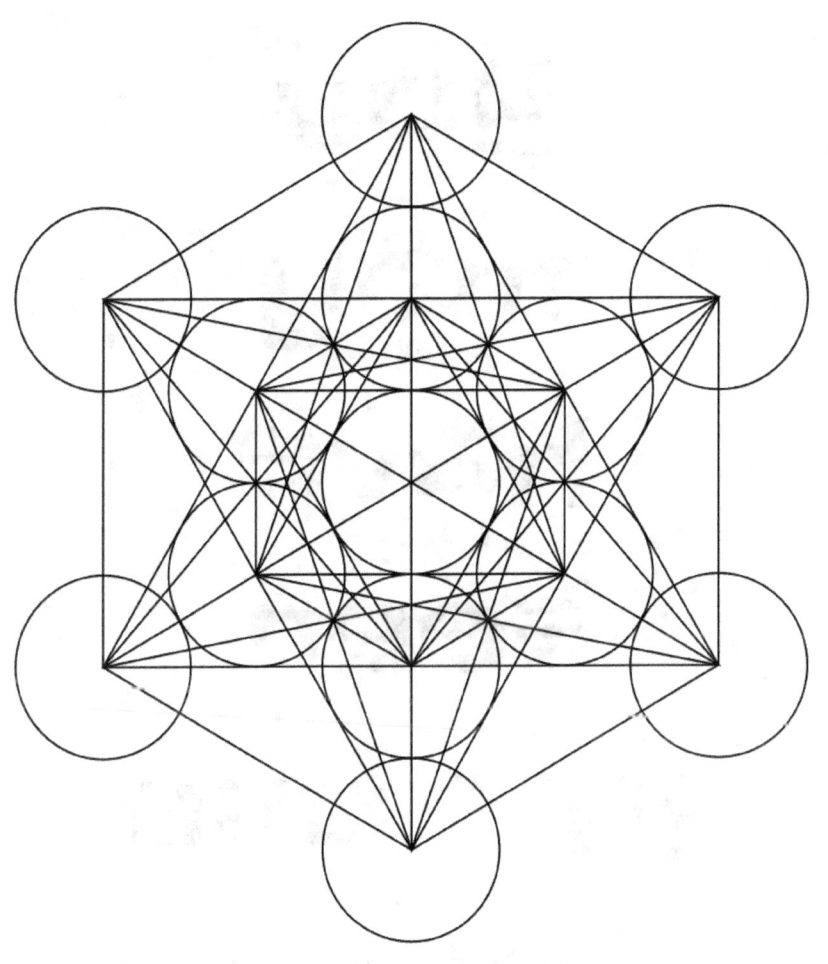

You may have seen this one before...

Snowflake

Energy
that is
frozen in
time
becomes
something
very
special.

We call this something... MATTER!

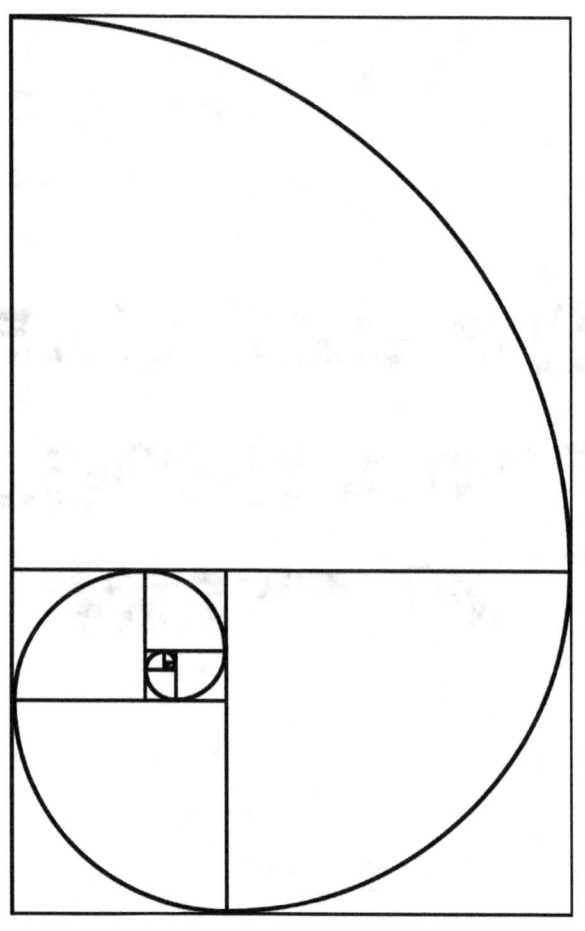

It is the
blueprint
for all the
things!

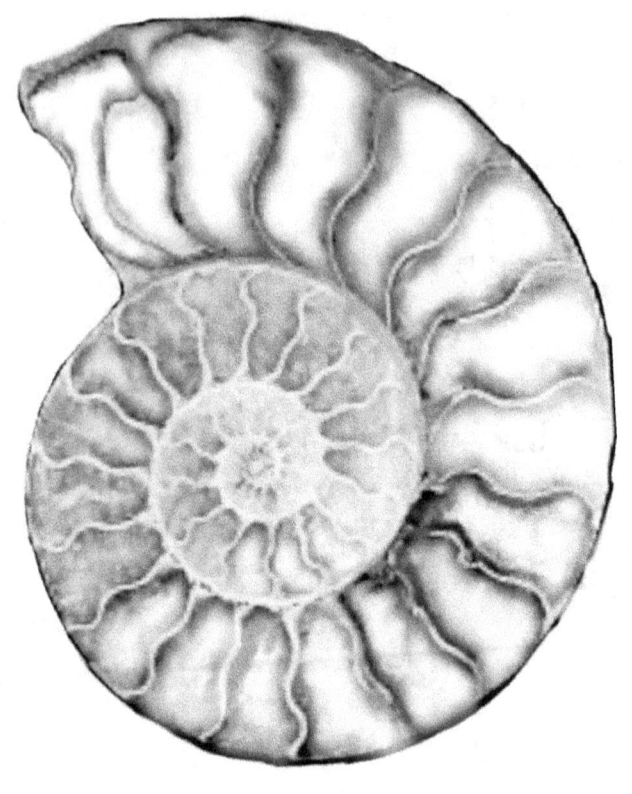

Seashell

Sunflower

pinecone

Grape

Pineapple

Leaf

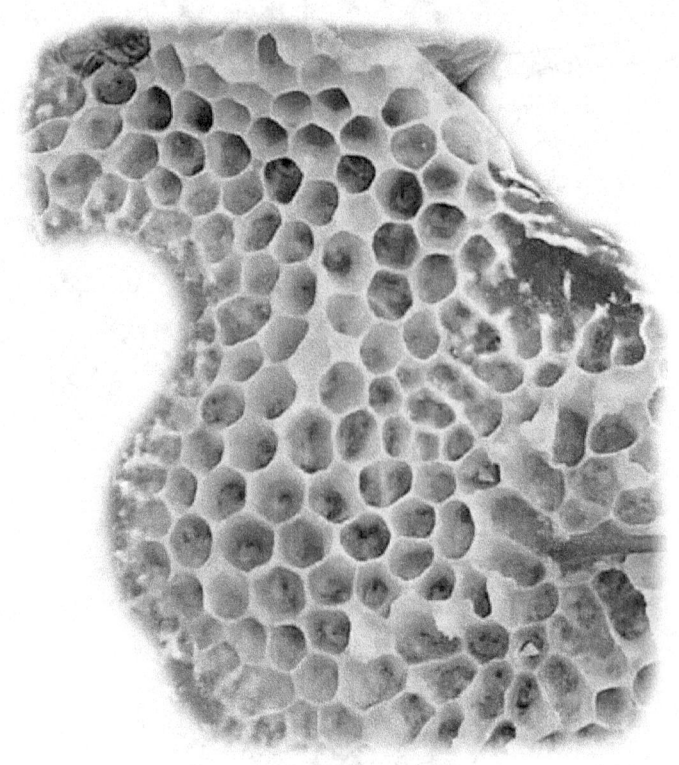

Honeycomb

Turtle

Rose

Giraffe

Tree

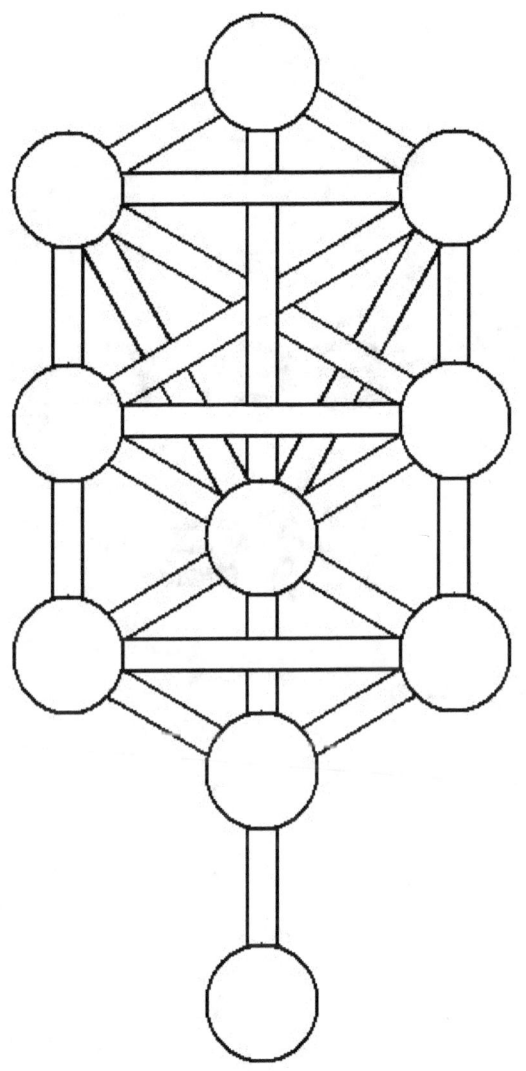

Tree of life

Energy is distilled into matter through the vibration of sounds.

ENERGY

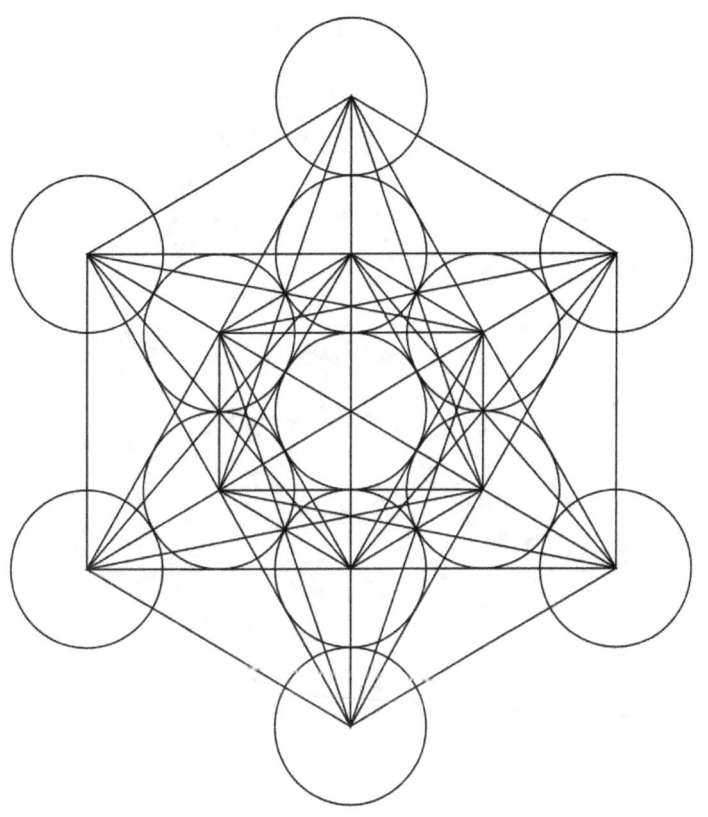

VIBRATION

MATTER

All that we are and all that we see begins and ends with Energy.

Now you know the Energy Secret!

The End.

www.ingramcontent.com/pod-product-compliance
Lightning Source LLC
Chambersburg PA
CBHW071829200526
45169CB00018B/1299